25¢

DESOTO SCHOOL LIBRARY

AF778250

DESOTO SCHOOL LIBRARY

DISCARD

The Discovery Books are prepared under the educational supervision of

Mary C. Austin, Ed.D.

Reading Specialist and

Lecturer on Education

Harvard University

A DISCOVERY BOOK

by *John Parlin*

illustrated by *Anthony D'Adamo*

GARRARD PUBLISHING COMPANY
CHAMPAIGN, ILLINOIS

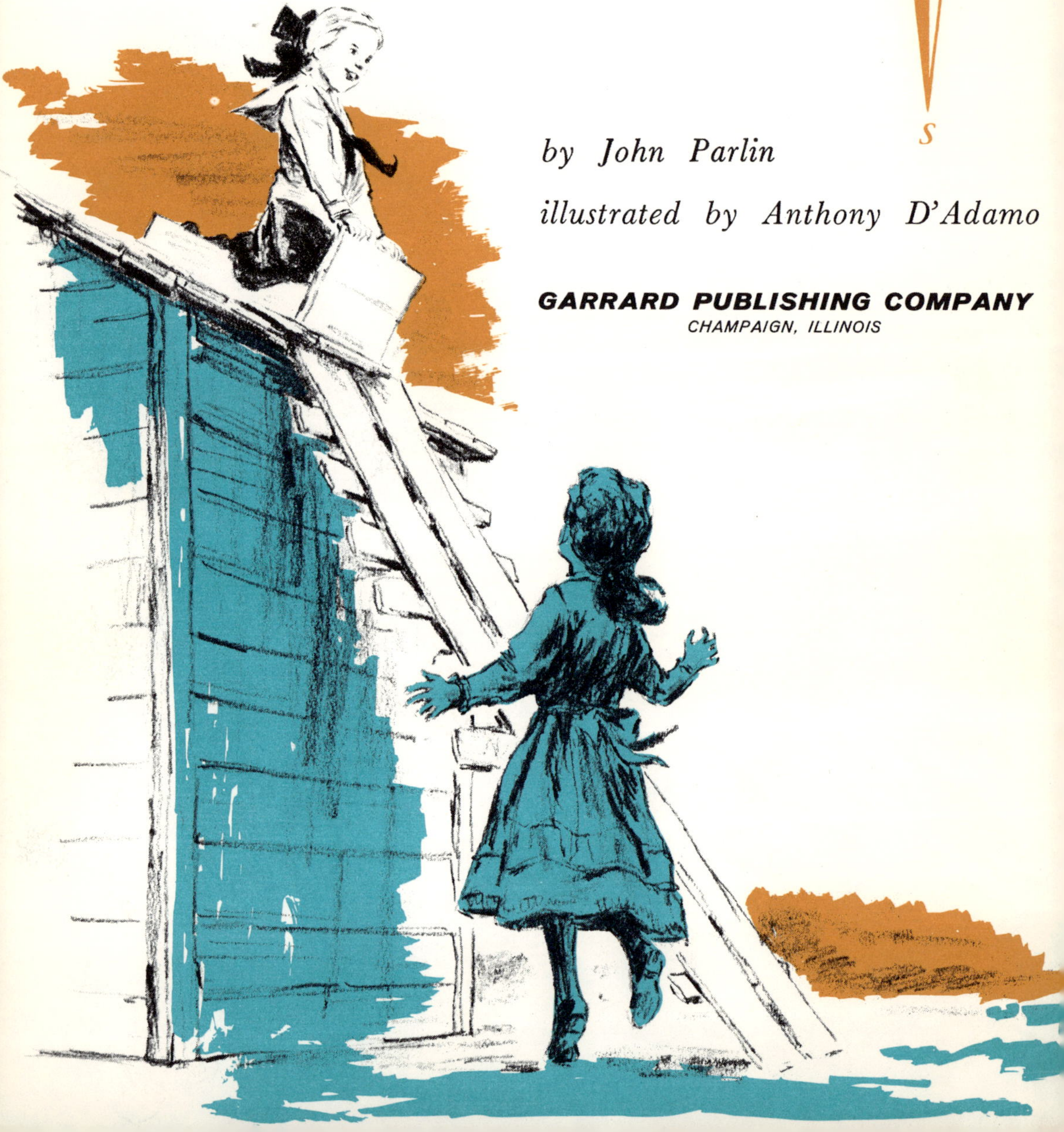

92
EAR

Amelia Earhart

Pioneer in the Sky

To those high-flying Parlin girls—

May, Alice and Alma

Copyright 1962 by John Parlin

Manufactured in the United States of America

Library of Congress Catalog Number: 62-8030

Contents

Amelia Earhart: Pioneer in the Sky

Chapter 1

First Airplane

"What will we do today?" Muriel Earhart asked her big sister, Amelia. Muriel was six and Amelia was eight.

"Let's make something," Amelia said.

"Make what?"

"Can't you think of something?" Amelia asked. "Do I always have to think of everything?"

"All right, smarty," Muriel said. "Let's make an airplane. Let's make it like the picture we saw in Daddy's magazine last night."

"I'm afraid we don't have the right things to make an airplane," Amelia said. "But I know what we can make that's just as exciting."

"What?" Muriel asked. "Tell me!"

"A roller coaster!" Amelia cried.

"We don't know how to make a roller coaster either," Muriel said.

"We can try. It won't hurt us to try."

The girls got some wood from an old fence that had been torn down. They made a "car" by putting some roller skates under a big piece of wood. With the rest of the wood they made a slide. The slide went from the top of the woodshed to the ground.

Amelia took the car up to the top of the shed. She put it on the slide and climbed in.

"Here I go!" Amelia shouted.

The car rolled quickly down the slide. When it hit the ground it bounced into the air. Amelia turned a somersault. Luckily she was not hurt.

When the girls' grandmother saw the roller coaster, she made them take it down at once.

"That's dangerous," she said. "What in the world made you want to ride on a thing like that?"

"Just for fun!" Amelia cried.

"You certainly have a strange idea of fun," her grandmother said.

Amelia and Muriel spent a great deal of time with their grandmother. Her house was in Atchison, Kansas. Amelia had been born there on July 24, 1898.

Their father worked for a railroad.

He had to travel a lot. But the girls had fun with him when he was home.

They had fun with their mother too. She let them bring frogs and spiders right into the house.

Later the family lived in Des Moines, Iowa. One day Mr. Earhart gave the girls a treat. He took them to the Iowa State Fair.

At the fair the girls rode on the Ferris wheel. When their car reached the top of the wheel, they screamed with delight. They could see for many miles. But as their car came down, Muriel cried with fear.

"There's nothing to be afraid of," Amelia said. "This is fun. It's almost like flying. Look! There's Daddy down on the ground."

The car reached the bottom, then swung toward the top again. Mr. Earhart waved to his girls. When the ride was over he asked, "How was it?"

"I was scared," Muriel said.

"I wasn't," Amelia bragged.

"Now let's see the airplane," Mr. Earhart said. He had never seen an airplane before. Neither had the girls. They had only seen pictures of airplanes in magazines.

As they walked across the fair grounds, Mr. Earhart talked about airplanes. He told them the Wright brothers had invented the airplane just a few years before. Airplanes were not able to fly very far yet.

Soon they came to a fence. On the other side they saw the airplane.

"It looks like a big orange crate," Amelia said.

"It looks funny," Muriel agreed.

"I thought it would look better than that," Amelia said.

She was disappointed by her first sight of an airplane. She never thought she would own an airplane when she grew up.

Chapter 2

Halley's Comet

One night in 1910, the Earharts were having supper. Mr. Earhart turned to Amelia and Muriel. "Would you girls like to see something exciting tonight?"

"Yes, Daddy, yes!" Amelia and Muriel shouted together. "What is it?"

"Well," Mr. Earhart said with a laugh, "it has a tail that is several million miles long."

"You're teasing us," Amelia said. "Nothing has a tail that long!"

"I'm not teasing," Mr. Earhart said. "I'm talking about Halley's Comet up in the sky." He told them that Halley's Comet made a trip near the earth every 76 years. It was named after Edmund Halley, an Englishman who had studied the comet long ago.

When the dishes were washed, the girls climbed up on a shed behind the house. Mr. Earhart pointed to Halley's Comet.

"It looks like a star with a tail," Amelia said.

"It's not a star," her father said. "We aren't sure exactly what it is. Most scientists think it's made up of millions of small rocks and pebbles."

"I read something interesting about comets the other day," Mrs. Earhart remarked. "Somebody said that a comet is the nearest thing to nothing that anything can be, and still be something."

"Well," said Muriel, "it looks like something to me."

"Isn't it pretty!" Amelia cried.

"Take a good look," Mr. Earhart said. "It will be out of sight in a few days. And it won't come back again until 1986."

Amelia pointed to a bright star near the Milky Way.

"Does that star have a name, Daddy?" she asked.

"That's Vega," her father replied. "It's one of the most beautiful stars of all."

Amelia shut her eyes and made a wish on Vega. She wished that when she grew up she could do exciting things.

When they went back in the house Amelia picked up the cat. "Well, Kitty Cat," she said, "Halley's Comet's tail is a lot longer than yours. You ought to be ashamed of yourself."

Of course, Amelia was just teasing. She really liked cats.

DESOTO SCHOOL LIBRARY

Chapter 3

Growing Up

The years that followed were happy ones for Amelia. She learned to swim well and to ride a horse.

After she finished high school she studied in Philadelphia. Muriel was going to a school in Toronto, Canada. One vacation Amelia went to Toronto to visit Muriel.

In Toronto Amelia saw many soldiers. The First World War was being fought in Europe. Canada had been in the war a long time. Many men had been wounded.

One day Amelia saw four soldiers walking on crutches. She felt sorry for the men. She decided to stay in Toronto.

"I can help these wounded soldiers," she told Muriel.

Amelia became a nurse's aid at a hospital. She brought medicine and food to the wounded men. The men liked Amelia.

On one of her days off, she and a friend went to an airfield near Toronto. They watched a man do some stunt flying. His daredevil tricks frightened Amelia's friend.

"I can't look," she cried, covering her eyes with her hands.

"Oh, I love it!" Amelia cried. She watched the plane do a somersault in the air. Next the man made his plane roll over. Then he spun down through the sky.

"It's wonderful!" Amelia shouted. "What fun it must be to fly!"

"I wouldn't dare fly," her friend said.

"I would," Amelia said quietly. "And someday I will."

It was several years before Amelia was able to fly. When the war was over, she went to Columbia University in New York City. She thought she might like to be a doctor. She felt that women could be just as good doctors as men.

Soon Amelia decided that she was not meant to be a doctor. But she could not decide what she wanted to be instead.

Her mother and father were now living in Los Angeles, California. So Amelia left Columbia University and went to California.

Chapter 4

First Flight

One morning in California Amelia picked up a newspaper. She read that there was going to be an air meet near Los Angeles. She begged her father to take her. Mr. Earhart liked to see airplanes fly too. So he agreed to go.

In those days flyers often had meets around landing fields. The meets were called "air shows" or "flying circuses." The pilots did lots of stunt flying.

Often a brave man jumped out of a plane with a parachute. He made a pretty sight as he floated to the ground.

Amelia was thrilled at the air meet. "I'd like to learn to fly," she said to her father. "I wonder how much it costs."

"Ask that man in uniform over there."

"You ask him for me, Dad," Amelia said. "Please! He might think a girl who wants to fly is crazy."

Amelia's father was back in a few minutes. "He said that it costs several hundred dollars!"

"That's a lot of money," Amelia said. "But, Dad, I'd . . . ."

"You might not like flying, Amelia," her father said hopefully. "You've never even been up."

"Maybe I should go up as a passenger at least once," Amelia said. "If I like it, I'm going to take lessons."

Mr. Earhart thought Amelia would be frightened on her first flight. But he was wrong.

Amelia loved her first flight. The planes in those days did not have glass windows or a roof. They were open to the wind and the rain and snow. Before Amelia went up she put on a flying helmet and some goggles. These were to protect her from the wind. She climbed up on the wing and stepped into the passenger seat.

A man on the ground spun the propeller. That's the way planes were started then. The engine started and the plane moved down the grassy field.

Faster and faster it went.

"Here I go," Amelia said to herself, repeating the words she had used on her roller coaster long ago.

The plane rose into the sky. A strong wind beat against Amelia's face. She stuck her arm out of the plane. The wind almost tore it off.

But Amelia had never loved anything quite so much. As soon as the plane was in the air, she made up her mind to learn to fly.

"I've got to," she said to herself, "no matter how much it costs."

Chapter 5

Flying Lessons

Amelia's father said he could not afford to pay for flying lessons. So Amelia got a job. She used the money she earned to pay her flying teacher.

Her flying teacher was a woman! Her name was Neta Snook. She was one of the few women in the world who knew how to fly.

Neta taught Amelia how to take off from the ground. She taught her how to bank when she made a turn. She also taught Amelia how to pull out of a dive when the plane stalled.

Finally, the day came for Amelia to solo, to go up by herself. Neta Snook had always gone up with her before. Now Amelia would be alone.

Amelia climbed into the plane. She raced it across the field and it rose into the air. Amelia felt like a bird.

But when she brought the plane back to the field she made a rough landing. Nobody said that she was a good pilot. But at least she had soloed. She had flown alone.

Amelia's mother was proud of her. She helped her buy a small plane.

Amelia soon became a good pilot.

"You take to flying like a baby takes to milk," someone said to her. Amelia liked that.

Planes did not fly very high in those days. But Amelia set a women's record for high flying. She flew to 14,000 feet. That's almost three miles high.

Amelia wanted to fly even higher. So she tried again. At 11,000 feet she ran into some thick clouds. There were sleet and snow in the clouds. They stung her face. She couldn't see.

She knew she had to get out of the clouds. So she went into a tail spin and dived for the ground. Finally she burst into clear sky beneath the clouds.

When she landed, a man asked her why she had come down in a tail spin.

"It was the fastest way," Amelia said.

"Suppose the clouds had reached all the way to the ground. You would have killed yourself," he said.

But Amelia was not worried. She loved flying more than ever.

"Nothing on sea or land can be more lovely than the realm of clouds," she said.

DESOTO SCHOOL LIBRARY

Chapter 6

Plans for an Ocean Flight

Amelia had a hard time deciding what she wanted to do. Of course, she wanted to fly. She wanted excitement. But she also wanted to do some useful work.

At last she found what she wanted. It was at Denison House in Boston. Denison House is a social center.

Amelia taught the children in the neighborhood how to play games. She looked after them while their mothers were working.

When Amelia wasn't taking care of the children, she was flying. The children were proud of her.

One day in 1927, Amelia read that Charles A. Lindbergh had flown an airplane from New York to Paris. He was the first man to fly the Atlantic Ocean alone. It was a great thing to do. Planes then were not nearly so good as planes today. The flight was long and dangerous.

Several women had tried to fly across the Atlantic Ocean. None had made it. Three young women lost their lives trying.

One day in April, 1928, Amelia was called to the telephone.

"Hello, Miss Earhart," a man said. "Would you like to make a long airplane flight?"

Amelia became excited at once. "Tell me more," she cried.

"I can't tell you over the phone," the man said. "It's a big secret. Come to my office and I'll explain everything."

When Amelia walked into his office he seemed pleased with her looks. She was tall and slender and appeared quite at ease. The man thought she looked like Lindbergh.

"How would you like to fly across the Atlantic Ocean?" he asked.

Amelia took a deep breath. "I'd like it," she said.

DESOTO SCHOOL LIBRARY

"If you are chosen, you'll probably just be a passenger," the man said.

"I'd like to be the pilot," she said.

"But you don't know how to use instruments. A man who does will be the pilot."

Instruments could help a pilot tell where he was at all times. They kept him from getting lost when it was dark or foggy. One of the most important instruments was a radio. Instrument flying was new. Amelia had never learned it.

Soon Amelia was sent to New York. There she met George Putnam, a book publisher. He was the manager of the ocean flight.

George and Amelia soon found that they liked many of the same things.

And they liked each other. George called Amelia "AE" after her initials. She began calling him "GP" after his initials.

Wilmer Stultz was picked to be the pilot on the ocean flight. Lou Gordon was the flight mechanic. Stultz was a fine pilot. He knew how to use instruments.

Amelia promised herself that someday she would learn instrument flying.

Chapter 7

Across the Atlantic

Amelia was pleased when she saw the plane that was chosen for the trip. It was named the *Friendship*.

The *Friendship* had three motors. Instead of wheels it had pontoons. This meant it could float. It had to take off and land on water.

"I like those gold wings," Amelia said. "They're beautiful!"

"We didn't choose gold because it's a pretty color," Wilmer Stultz laughed.

FRIENDSHIP

"No," Lou Gordon said. "That color can be seen a long way off. If we have to come down in the ocean, perhaps we can be rescued."

The men couldn't scare Amelia. She felt sure the *Friendship* would make the flight safely.

In 1928, planes did not fly well in bad weather. It was hard to leave the ground unless the wind was just right.

Early one morning the weather in Boston seemed right. Amelia and the men climbed aboard the *Friendship*. It was in the waters of Boston Harbor.

The engines were started. Seagulls screamed as they heard the roaring motors. The *Friendship* sped across the water. Foam flew like soapsuds beneath it. Slowly the plane rose into the air.

They flew to Trepassy Harbor, in Newfoundland, which is off the coast of Canada. There they came down in the water. They planned to fill up again with gasoline, then take off for Europe.

But the weather was bad. Day after day they had fog and rain. Sometimes they tried to take off, but the wind was wrong.

Amelia became discouraged. There was some bad news too. Another young woman had flown from New York to Newfoundland. She and her pilot were in a town nearby. She hoped to beat Amelia across the Atlantic.

Her name was Mabel Boll. She was sometimes called "The Diamond Queen" because she wore so much jewelry. Her plane was named the *Columbia.*

Amelia said it was not a race. And Mabel agreed with her. But of course, each wanted to be the first woman to cross the Atlantic by plane.

Back in the United States the newspapers called it a race. Headlines read:

"Rival Women Flyers Still
Held by Bad Weather"

"*Columbia* Ready for Take-Off"

"Weather Delays Both Planes"

and finally, in big type,

"*FRIENDSHIP* TAKES OFF"

It was June 17, 1928. Mabel Boll said the weather was too bad to fly. So if the *Friendship* made it, Amelia would be the first woman to fly the Atlantic.

At first, everything went fine. But soon Amelia and the men ran into heavy fog.

They had hoped to reach Southampton on the southern coast of England. But after many hours in the air they had little gasoline left. If they ran out of gasoline, they might crash into the sea.

Then the radio stopped working. Without the radio they could not tell where they were.

They were lucky. The fog lifted. They saw a big ship below. Stultz circled the ship. Then he wrote a note. He asked the Captain to paint a big sign on the deck. The sign should tell the flyers where the nearest land was.

Stultz gave the note to Amelia. She tied it in a bag with an orange to weight it down. Then she dropped the bag out of the plane. She hoped it would land on the ship.

FRIENDSHIP

It missed! Amelia tried again with another note. She used their last orange as a weight.

Amelia missed again!

They decided to go on. There was still a chance of finding land before their gasoline gave out.

A short time later they saw some small fishing boats. They knew that there must be land nearby. But where could it be?

Something blue appeared in the distance. It could be land. Or it might be just another cloud.

When they came nearer, they saw that it really was land!

Stultz brought the plane down on the water. They were just off Burrey Port, Wales. Wales and England are both part of Great Britain.

Amelia was the first woman to fly the Atlantic. She was treated like a heroine. Crowds cheered her wherever she went.

The President of the United States, Calvin Coolidge, sent her a message. Mabel Boll sent a nice message too.

Many people compared Amelia to Lindbergh. They called her "Lady Lindy." Amelia knew she did not deserve to be called "Lady Lindy."

"I was just a passenger," she said. "I might as well have been a sack of potatoes."

But saying that just made people like her even more. They admired her modesty.

"Someday," Amelia said to herself, "I'm going to fly the Atlantic all by myself—solo!"

Chapter 8

Another Wish Upon a Star

Amelia and the two men returned to America. The country gave them a big welcome. There was a parade up Broadway in New York City.

George Putnam made Amelia write a book about the flight. He wanted it for his publishing company. Of course Amelia would make money from the book.

The book was called *20 Hrs. 40 Min.* That's how long it took the *Friendship* to fly across the ocean.

"It takes a lot longer than 20 hours and 40 minutes to write a book," said Amelia. But GP helped her. He was a good writer.

When the book was finished, Amelia began flying again. She wanted to get more practice as a pilot. One day she started across the country to California in a small plane.

It was a long flight in those days. She had to stop many times for gasoline. She could not fly at night. Most of the landing fields were not lighted. And there were not many landing fields.

Once she became lost. It was almost dark. Her plane was nearly out of gasoline.

Amelia could find no place to land. But she knew she had to land very soon.

She circled low over a small town. It had a wide main street. There were no cars on the street.

"I'll have to try it," Amelia said to herself. She made a fine landing on the main street.

The people who lived there were very surprised! They had never had a visitor from the skies before. They were proud when they learned the flyer was the famous Amelia Earhart.

"Flying may not be all plain sailing," Amelia said later, *"but the fun of it is worth the price."*

Amelia returned to New York. She got a job on *Cosmopolitan Magazine.* She wrote about flying. She told her readers that many children knew much more about airplanes than grownups.

She called the children "the flying generation."

Often young people who wanted to learn to fly wrote to Amelia. Many said their parents did not want them to learn. Some told Amelia they took lessons secretly.

Amelia thought that was bad. She said that parents should help their children learn to fly. "The parents should be sure their children learn the *right* way," Amelia said.

When Amelia left the magazine, she worked for a new passenger airline. Her job was to make women like flying. Some women thought flying was not safe. They did not want their husbands to travel by airplane. Amelia made many women think differently about flying.

She also flew a great deal herself. She was becoming a better pilot than ever before. And she was learning to fly with instruments.

Amelia bought a plane that was called a "Vega." She remembered the time when she had first seen the star named "Vega."

"What a good name for an airplane," she thought. "Maybe I should make another wish."

Amelia wished that she would be the first woman to pilot a plane across the Atlantic—alone!

DESOTO SCHOOL LIBRARY

DESOTO CO. SCHOOL LIBRARY

Chapter 9

Atlantic Solo!

"Will you marry me?" George Putnam asked Amelia one day.

"Marriage and flying don't go together," Amelia said.

"But I'll help you with your flying," George said. He finally changed her mind and they were married.

One day Amelia said to her husband, "I'd like to fly the Atlantic Ocean alone."

George knew the trip would be dangerous. But he had promised to help Amelia with her flying. "If you want to do it," he said, "I'll help you all I can."

Amelia bought a new Vega. When everything was ready she flew to Newfoundland. Just after 7 p.m. on May 20, 1932, she took off for Europe —alone.

While Amelia was over the Atlantic, Americans held their breath. Would she make it? Everywhere people prayed that she would.

Amelia did not have an easy flight. Ice formed on her plane's wings. She had to come down close to the sea where the air was warmer. When the ice melted she climbed again.

Suddenly she saw flames coming from the exhaust pipe. It could mean trouble. The plane might catch on fire.

Then a gasoline tank started leaking. It would be terrible if the gasoline caught fire from the exhaust flames.

"I'd rather drown than burn up," Amelia said to herself. So she brought the plane down close to the waves again.

The flames did not seem to get any bigger. So Amelia felt safer. Her plane roared on and on through the night.

The sky slowly grew light. Amelia was hungry. She drank some hot chicken soup from a thermos bottle and ate two chocolate bars. Then she made a hole in a can of tomato juice and drank it.

Amelia had hoped to fly all the way to Paris. But she decided not to risk it.

The flaming exhaust and the leaking tank made her change her plans.

She was certainly glad when she saw land ahead. She knew it was Ireland. At last she had flown the Atlantic alone. Solo!

Amelia brought her plane down in a field. It had taken her just 15 hours and 18 minutes to fly across the Atlantic. It was a new record.

Soon after she landed, Amelia went to London. A newspaper said, *"Not America only, not women only, but the whole world is proud of her."*

Amelia was invited to many parties. One night she danced with the Prince of Wales. He would be the King of England when his father died. They talked about flying while they danced.

WELCOME TO
NEW YORK
LADY LINDY

Afterwards someone asked Amelia if the Prince was a good dancer. *"The prince is a flyer and all flyers dance well!"* Amelia said.

George Putnam came over by ship to meet Amelia. Together they went to France, Italy and Belgium. In Belgium they had lunch with the King and Queen. Afterwards the King gave Amelia a medal.

When Amelia returned to New York thousands of people cheered her. She received many honors.

But Amelia wanted to give some of the credit for her flight to her husband. *"It was much harder for him to stay behind than it was for me to go,"* she said.

Chapter 10

American Heroine

Amelia and George were invited to have dinner with President and Mrs. Herbert Hoover at the White House. Afterwards they went to Constitution Hall where the President gave her a medal. It was from the National Geographic Society.

"The nation is proud," President Hoover said, *"that an American woman should be the first woman in history to fly an airplane alone across the Atlantic Ocean."*

Amelia had proved that women have as much courage as men. She was now one of the most famous women alive.

Amelia still liked cats. She was pleased when she was asked to present a kitten to the Explorers Club in New York City. The kitten, a blue Persian, was to be the Club's mascot.

She held the kitten in her arms as she talked to the club members. She told them that there was an old saying about kittens. If you oiled their feet they would stay home. Perhaps people thought their feet would be too slippery to run away on.

So Amelia rubbed oil on the kitten's feet. Then she gave it to the explorers. They named it "Amelia."

Amelia was making many new friends.

She was invited to dinner again at the White House. Her hostess was Mrs. Franklin D. Roosevelt, the new President's wife.

"Would you like to go up in a plane after dinner?" Amelia asked Mrs. Roosevelt.

"I certainly would," Mrs. Roosevelt said. "I've never flown at night."

Amelia arranged for the pilot of a big plane to take them up. George Putnam and some newspaper reporters went with them.

When they were high over Washington Amelia and the others looked down. They saw the lighted dome of the Capitol building glowing in the darkness. The white shaft of the Washington Monument shone in all its glory.

Amelia went up to the cockpit to speak to the pilot. She had on an evening gown and wore white gloves and slippers. She asked the pilot to let her fly the plane. He gave Amelia his seat.

Back in the cabin a reporter asked Mrs. Roosevelt, "Do you feel just as safe knowing a girl is flying this ship?"

"*Just as safe,*" Mrs. Roosevelt replied. "*I'd give a lot to do it myself.*"

Mrs. Roosevelt said that it really marked a new day in history "*when a girl in evening dress and slippers can pilot a plane at night.*"

Chapter 11

"A Challenge to Others"

Amelia became a teacher at Purdue University in Indiana. She taught aviation. The students called her "The Flying Professor."

Purdue gave Amelia an airplane. It was a Lockheed Electra, much bigger than any plane she had ever had.

One night Amelia and George were talking at home. Amelia pointed to a globe map.

"I want to fly all the way around the world," she said.

"Other pilots have done it," George said.

"I know," Amelia said. Then she pointed to the equator on the middle of the globe. "But no one has ever flown around the world at the equator. That's the longest way around."

"And it's the hardest way around," George said.

"Will you help me make my plans?" Amelia asked.

George remembered his promise. "Yes," he said, "you know I will."

One morning in March, 1937, Amelia took off from California. Fred J. Noonan was her navigator.

Their first stop was Honolulu, Hawaii. They made the flight in less than 16 hours.

The next day they climbed into the Electra. As the plane raced down the runway one of the wings dipped. Amelia tried to right the plane. But she was helpless. The Electra went into a ground loop.

Nobody was hurt. But the plane was a wreck. It had cost $80,000. It would cost a lot of money to fix it. "But I'm going to have it fixed," Amelia said. "I'm going to try again."

The plane was brought back to California by ship. While it was being repaired, Amelia and Fred Noonan made new plans.

They decided to try to fly around the world in the opposite direction. They would still start in California but they would fly east.

Amelia received hundreds of letters. Many were from boys and girls who wanted to go on the trip.

"I want to see the world," one boy wrote. "I have no money, but I will work my head off." Amelia was sorry she could not take him with her on the long flight.

When the plane was ready Amelia and Fred flew to Miami, Florida. Here the last plans were made. George Putnam arranged for the right kind of gasoline to be at every place they planned to stop.

On June 1, 1937, Amelia and Fred started around the world again. They flew to South America. They crossed the Atlantic Ocean to Africa. Then they went on to India.

After many days they reached Lae, New Guinea. New Guinea is an island in the Pacific Ocean far across from America.

Amelia and Fred planned to fly to Howland Island next. Howland Island is a tiny speck of land in the middle of the Pacific Ocean.

They knew it would be hard to find. If they came down in the ocean they would probably drown. But they must stop somewhere for more gasoline. And gasoline was waiting for them there.

The United States Coast Guard sent a ship to Howland Island. It was named the *Itasca.* It had a strong radio that could send messages. Amelia hoped the *Itasca's* radio could lead her safely to Howland.

On July 2, 1937, Amelia and Fred took off from Lae. They climbed into the sky and flew above the blue ocean.

The *Itasca* tried to reach them by radio. But they were still too far away.

"Have not heard your signals yet," the *Itasca* radioed. "Go ahead. Am listening now."

Amelia did not answer. She probably did not hear the message.

Finally, Amelia's voice was heard. But there was a lot of noise in the radio. The *Itasca* could not understand what she was saying.

Suddenly, Amelia was heard again. She said she was about 100 miles from Howland Island and the ship. She was almost there.

The *Itasca* kept on sending messages.

Amelia broke in. She sounded excited and frightened. *"We must be on you, but cannot see you,"* she cried. *"But gas is running low."*

It was clear to the men on the *Itasca* that Amelia was lost. If she had been where she thought she was, she could have seen the ship.

Then the *Itasca* heard Amelia say, *"We are running north and south."*

Those were the last words the *Itasca* heard Amelia speak. Amelia never did find the *Itasca* or Howland Island. Her plane must have come down at sea when it ran out of gasoline.

The United States Navy sent many ships and planes to search for Amelia Earhart and Fred Noonan. But they could not find them.

No one knows for sure what happened to Amelia and Fred. Since 1937, there have been many strange stories about their fate. Some people think perhaps the Japanese Navy captured them.

All we know for sure is that they disappeared. They probably had to come down in the ocean and were drowned.

Before Amelia left on her dangerous flight, she wrote a letter to her husband. She said that he was to read it only if she were lost.

When all hope for Amelia was gone, George Putnam read the letter. It said:

"I want to do it because I want to do it. Women must try to do things as men have tried. When they fail, their failure must be but a challenge to others."

Amelia's failure was a challenge to others, and a help too. She was a pioneer in the sky. The things she learned about flying helped the flyers who came after her. Partly because of the work of Amelia Earhart and Fred Noonan, planes now fly safely over the oceans every day.

DESOTO SCHOOL LIBRARY

DATE DUE

APR 26 '71			
MAY 5 '71			
JUL 8 '71			
JUL 8 '71			
OCT 13 '72			
MAR 9 '73			
DEC 20 '76			
MAY 08 1992			
APR 1 1993			
APR 8 1993			

170 IDEAL PRINTED IN U.S.A.